EGYPTIAN GENIUS

CONTENTS

JAMES BURNETT
CALVIN IRONS

Ancient Egypt

Pharaoh Tutankhamun funeral mask.

Egypt is one of the most famous places in the world. Egyptian civilization began thousands of years ago, and many great developments and discoveries can be traced back to those ancient times.

Egypt is part of the continent of Africa. Most of the country is stony desert, but there is rich land along the river Nile.

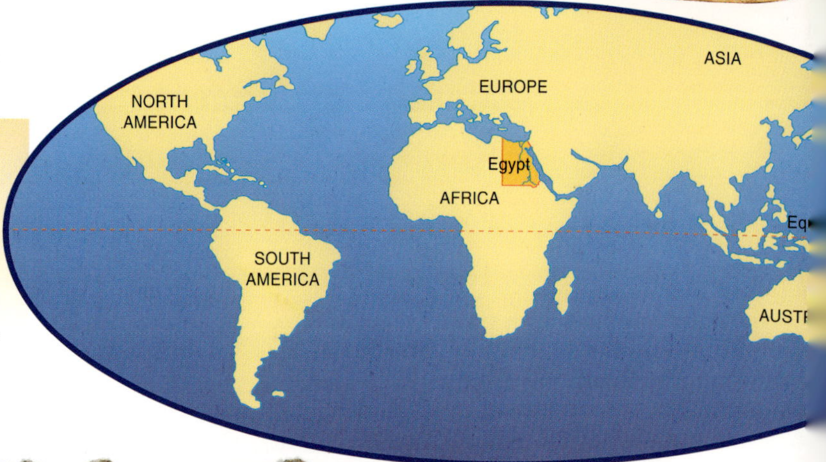

The Nile river ofter featured in Egyptia art. This tomb painting shows peo hunting on the Nil

The timeline below shows approximate dates for some events in Egypt's history. Some of the buildings would have taken many years to complete.

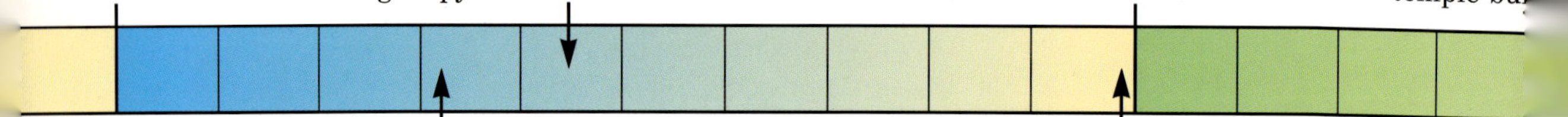

Great Pyramids at Giza
There are three pyramids at Giza. Each is hundreds of feet tall.

Queen Hatshepsut's Temple
Queen Hatshepsut was one of ancient Egypt's rulers. A huge terraced temple was built in her honor.

Temples of King Ramses II at Abu Simbel
Ramses' temples were carved out of sandstone cliffs.

Temples at Karnak
The temples at Karnak are taller than a nine-story building – and the columns are still standing today.

Alexandria was the site of the Pharos Lighthouse – one of the seven wonders of the ancient world. Standing more than 400 feet tall, it guided ships for many years before being destroyed by an earthquake in the 1300s.

A World of Activity

The Nile river was at the center of all Egyptian life. Many great buildings were constructed along its banks, and it provided a means of transport and a place for sport. The Nile also supplied the water and fertile soil that made successful farming possible.

Look at all the pictures on pages 2 – 5.

1. When do you think the Egyptians would have needed:

a. to measure things?

b. to use numbers?

2. What other mathematics do you think the ancient Egyptians would have used?

Look at the timeline on pages 2 – 3.

3. Estimate when:

a. the largest pyramid at Giza was built.

b. Queen Hatshepsut's temple was built.

4. What happened about:

a. 4,000 years ago?

b. 4,700 years ago?

5. About how many years ago did the Romans invade Egypt?

About the Nile

- The Nile is the world's longest river: 4,145 miles
- It flows from south to no
- About 1,108,000 square miles of land drains into the Nile. This is about one tenth of Africa.

The River's Cycles

The ancient Egyptians developed a calendar that was based on the three "seasons" of the river: the time of flood; the time when the water level fell; and the time of drought.

A device called a shaduf was used to lift water out of the river and into irrigation canals.

MASSIVE MONUMENTS

The great stone pyramids of Egypt are probably the most famous buildings in the world. Each one was built as a tomb for an Egyptian king or *pharaoh*. Building a pyramid took up to 20 years, and involved thousands of workers.

Not all pyramids were built to the same design.

Step Pyramid

The first pyramid was a step pyramid, built around 2680 B.C. for King Djoser. The six steps are like box shapes stacked one on top of the other.

Bent Pyramid

The Egyptians called this bent pyramid "the Gleaming Pyramid of the South." The bottom half of the pyramid has steeper sides than the top half.

True Pyramid

In the design known as the "true pyramid," each side follows a straight line to the top.

The Pyramids at Giza

More than 80 pyramids still stand along the west bank of the Nile river. The most famous are the great pyramids at Giza. These are the tombs of three pharaohs: Khufu, Khafra, and Menkaura.

Architects and mathematicians played key roles in the planning of the pyramids. Their calculations were essential in ensuring that every block of stone was cut and positioned correctly. The overall position of a pyramid was calculated carefully as well. Each of the great pyramids at Giza was built so that its sides face directly north, south, east, and west.

King Khufu's Pyramid

King Khufu's tomb is the largest of the three great pyramids at Giza. It once stood 485 feet tall – about as high as a modern 40 story building. Over the years, a section of about 33 feet at the top has crumbled away.

GIANT BUILDING BLOCKS

Khufu's pyramid was built using more than 2,000,000 huge blocks of limestone. Most blocks weighed about 5,500 pounds. Some of the biggest blocks weighed more than 100,000 pounds.

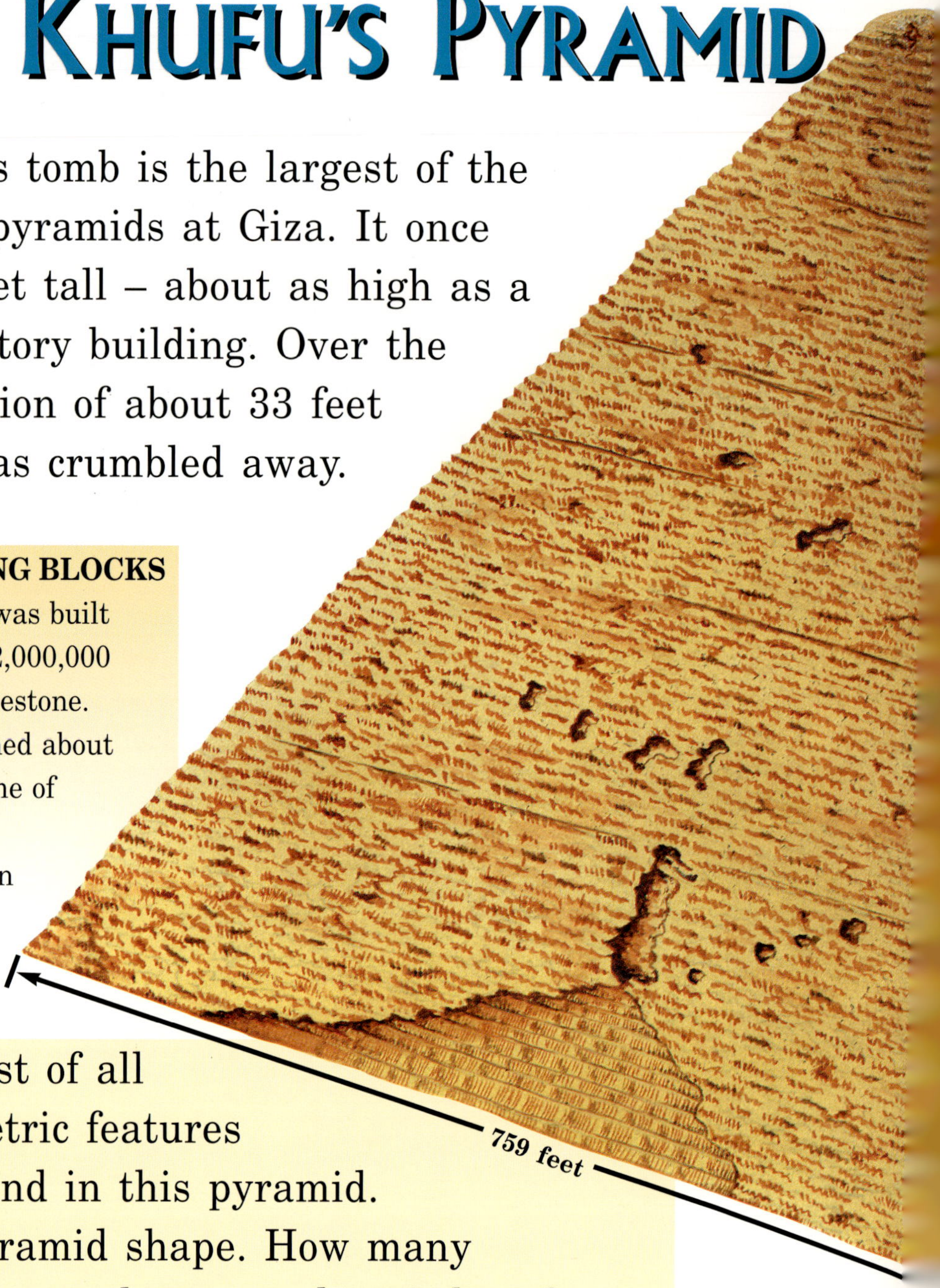

1. Make a list of all the geometric features you can find in this pyramid.

2. Find a pyramid shape. How many faces, edges, and corners does it have? Draw a picture of your pyramid.

3. Khufu's pyramid has a square base. What would the perimeter of the base be?

4. Look at the other famous buildings shown on page 9. Estimate the height of each one.

HOW DO THEY MEASURE UP?

Khufu's pyramid is taller than some of the world's famous statues, buildings, and towers.

Research

- About how many football fields would fit in a square as big as the base of Khufu's pyramid?
- Find out the dimensions of the other two pyramids at Giza. How do these compare with Khufu's pyramid?
- Find out the height of a tall building or tower you know. How does this compare with the height of Khufu's pyramid?

PYRAMID PATTERNS

Not all pyramid shapes look the same. They all have some triangular faces that meet at one point, but they don't all have the same-shaped base.

Make a Chart

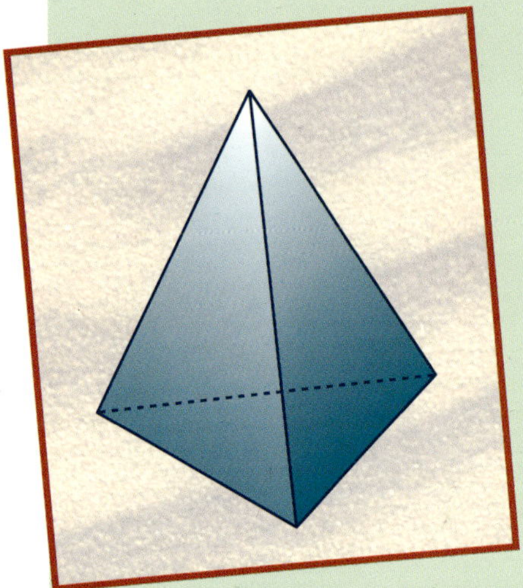

Look at each of these pyramids.

- How many sides does the base have?
- How many flat faces does the pyramid have altogether?
- How many vertices does it have?
- How many edges does it have?

Construct a chart to organize your data for all four pyramids.

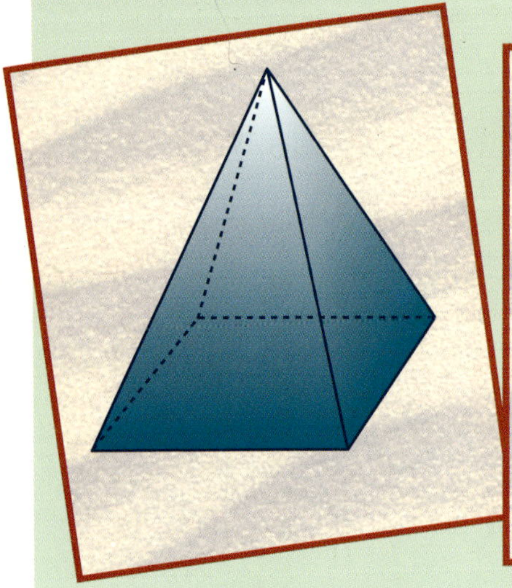

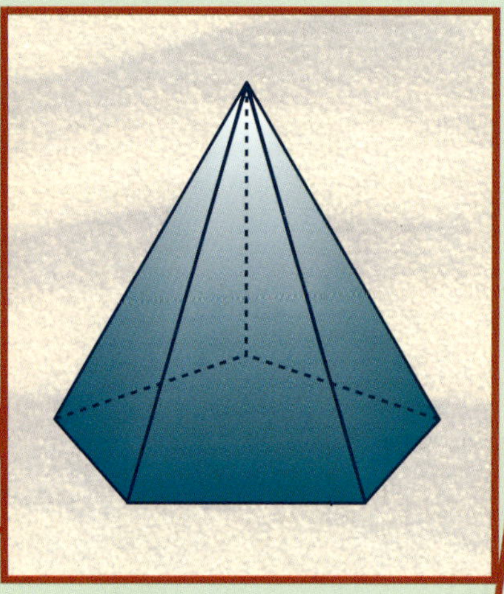

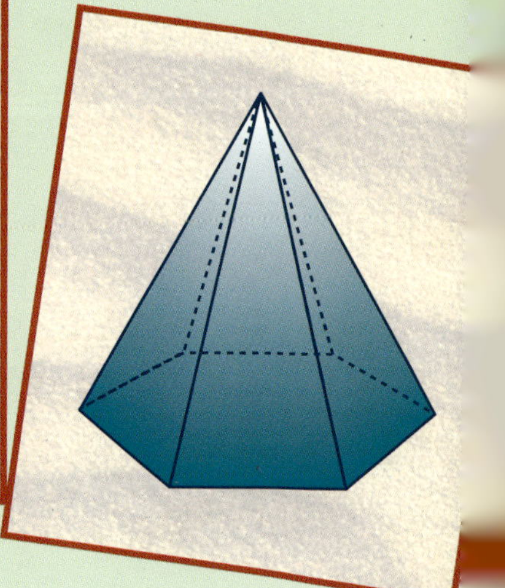

Look at the chart you constructed.

1. What pattern do you notice when you look at the number of faces of all the shapes?

2. What pattern do you notice when you look at the number of edges of all the shapes?

3. What other patterns can you see?

4. Suppose a pyramid had a base with seven sides. Predict how many faces, edges, and vertices it would have.

A "Pyramidal" Number

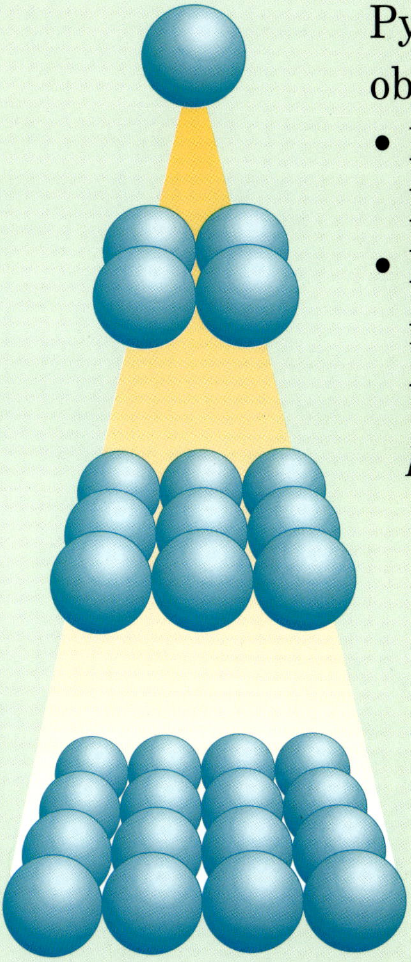

Pyramids can be made by stacking objects on top of one another.

- Look at the diagram at left. How many balls are in each layer?
- How many balls altogether are in the stack below? Why do you think this number is called a *pyramidal number*?

Tomb Paintings

Many Egyptian tombs were decorated with paintings. The pictures showed events from the person's life, and hopes and ideas about what would happen after his or her death. These tomb paintings were planned and completed with great care.

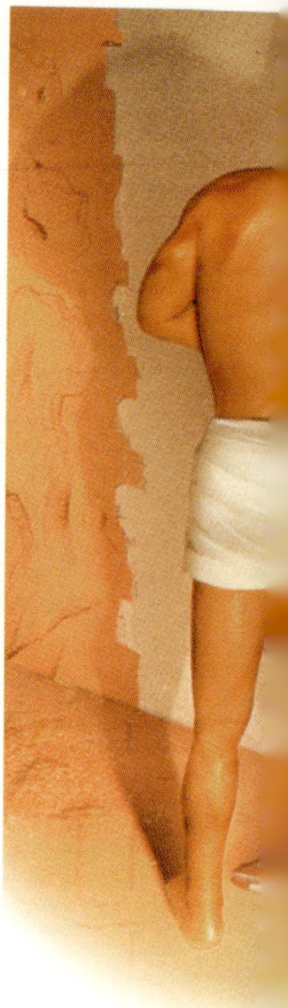

An Artist's Sketches

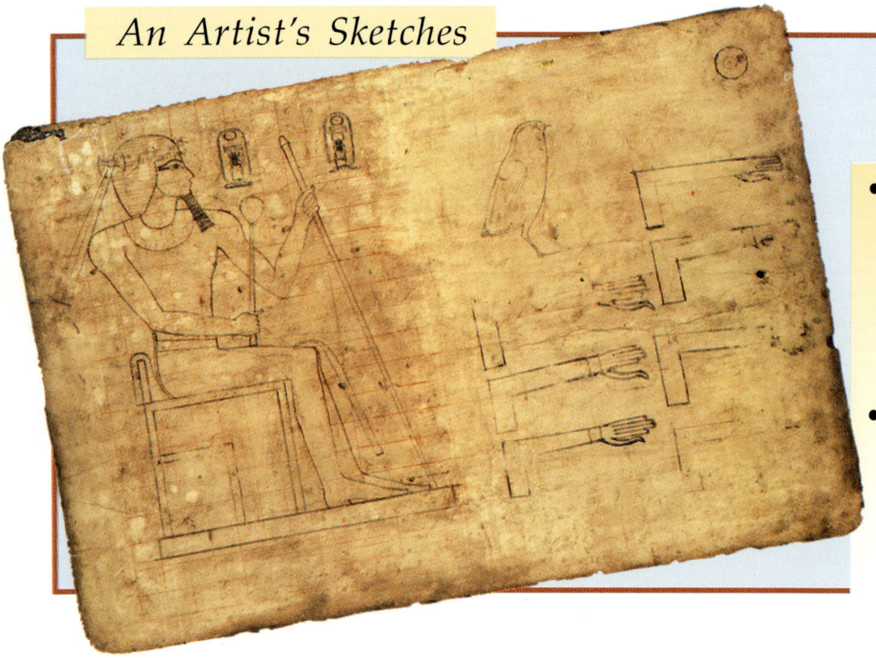

- There were at least three stages in creating a tomb painting: marking a grid; outlining the drawing; and applying color.
- Different artists worked on each stage. This "sketchpad" shows a practice drawing by an outline artist.

Marking the Grid

String coated in red paint was used to mark a tomb wall with a grid of squares. This grid was an important guide for the artist who worked on the outline drawing.

Artists always showed people's features in the same way. Chests and shoulders were shown from the front, but faces and legs from the side. Both big toes were put on the same side of the feet.

HAND DRAWN

Artists working on tomb paintings were guided by certain "rules," especially when they were drawing the main person. Each section of the body had to measure a set number of grid squares.

Drawing by the Rules

- The palm of the hand was as wide as one grid square. So on a grid of two-inch squares, the palm would be two inches wide. On a grid of two-foot squares, it would be two feet wide.
- The height from the bottom of the feet to the top of the head was 18 squares.
- The arms were one square wide at their thickest point.
- The eye was $\frac{3}{4}$ of a square long.
- The leg from the top of the knee to the bottom of the foot was six squares.

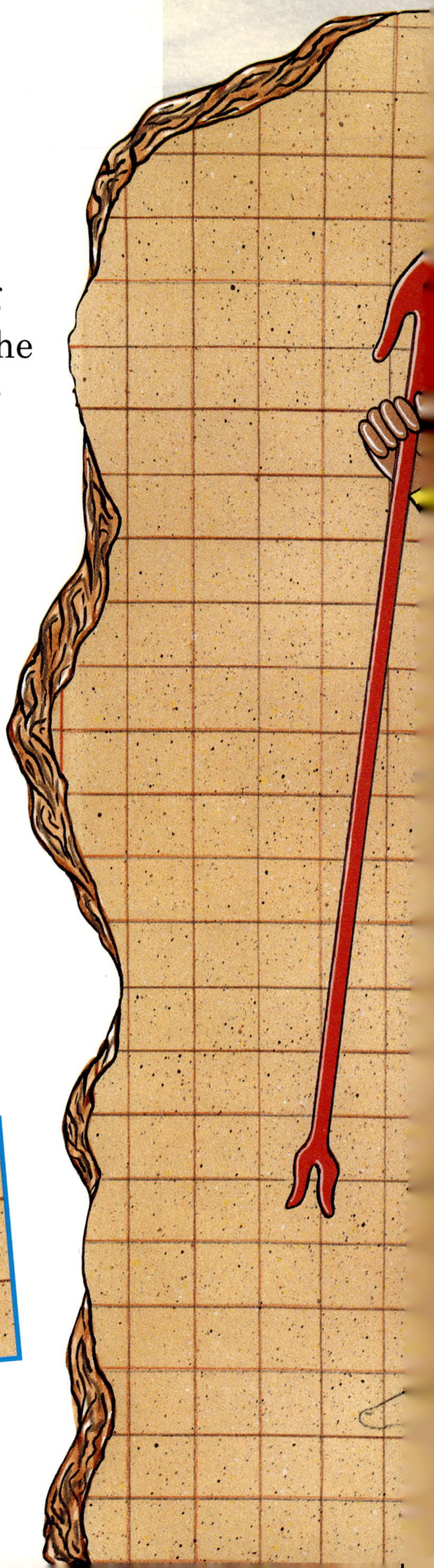

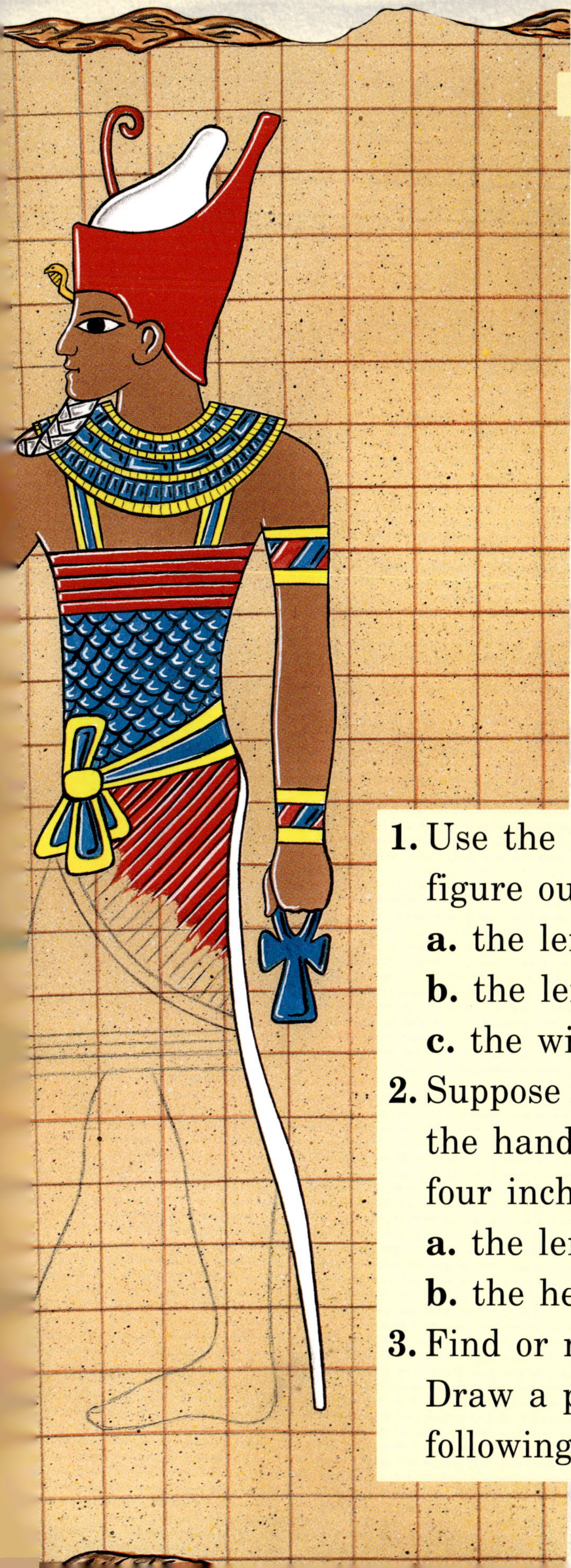

How do you compare?

Measure the width of your hand. Following the Egyptian rules, predict:

a. the width of your arm

b. the distance between your knee and the bottom of your foot

c. your height.

Now measure these dimensions of your body. How close were your predictions?

1. Use the illustration to help you figure out the rules for:
   a. the length of an arm
   b. the length of an ear
   c. the width of the shoulders.
2. Suppose the width of the palm of the hand in the illustration is four inches. Calculate:
   a. the length of the eye
   b. the height of the figure.
3. Find or make a grid of squares. Draw a person on the grid, following the Egyptian rules.

It's About Time

The ancient Egyptians developed many ways of measuring time. They had a calendar with 12 months, like the modern calendar, and they were the first people to calculate that one year is about $365\frac{1}{4}$ days long.

An Egyptian Calendar

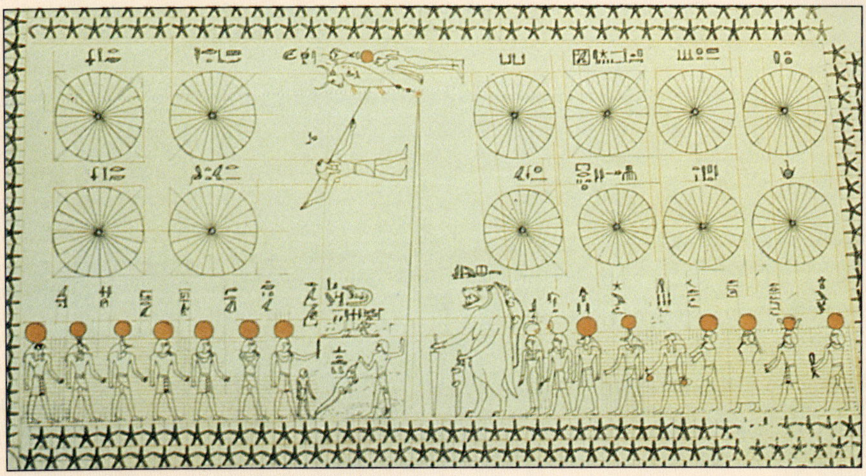

A full cycle of the moon (from full moon to full moon) takes a little less than 30 days. The Egyptians based their months on this cycle; they divided their year into 12 months of 30 days each, with 5 *holy* days at the beginning of each year. This calendar shows the 12 months of the year as circles.

The great columns of this temple helped the Egyptian to measure a year. They were positioned so that the rising sun would shine directly between them only on the longest day of summer. By counting days, the ancient Egyptians determined that this happened only once every $365\frac{1}{4}$ days.

The Temple of Karnak

For a long time, the early Romans had an inaccurate calendar, because it was based on the idea that a year had only 355 days. Finally the emperor Julius Caesar asked an Egyptian to help, and the year 46 B.C. was "stretched" to 445 days to correct the problem.

What is a year?

A year is the length of time it takes the earth to revolve once around the sun. The exact length of a year is 365 days, 5 hours, 48 minutes, and 46 seconds.

What is a month?

What we call a month is based on a *lunar month*. A lunar month is the length of time it takes the moon to revolve once around the earth. One complete cycle is about $29\frac{1}{2}$ days long.

FROM TIME TO TIME

The ancient Egyptians had several ways of keeping time. Their first "clocks" used shadows. With shadow clocks, the period of sunlight each day was divided into 12 equal parts. When days stayed light for a long time, these "hours" were longer than at times when night fell early.

Read about the shadow clock.

1. a. Why do you think the shadow clock had to be turned around at noon?

b. How many hours after noon would the shadow have been the longest?

2. Suppose an Egyptian "hour" was 45 minutes long. How many minutes of sunlight would there have been:

a. before noon?

b. from sunrise to sunset?

3. List some advantages and disadvantages of:

a. the water clock

b. the shadow clock.

A water clock

The Egyptian water cloc was a later developmen than the shadow clock. A water clock involved two stone bowls, one of which emptied into the other. As the water flowed out, markings on the inside of the emptying bowl became visible. Each new marking showed that o "hour" had passed.

A Shadow Clock

A shadow clock relied on the sun. The upright T-shaped piece created a shadow that fell along the horizontal bar. The first shadow was the longest, and marked the sixth hour before noon. Noon had the shortest shadow. Then the shadow clock had to be turned around for the afternoon.

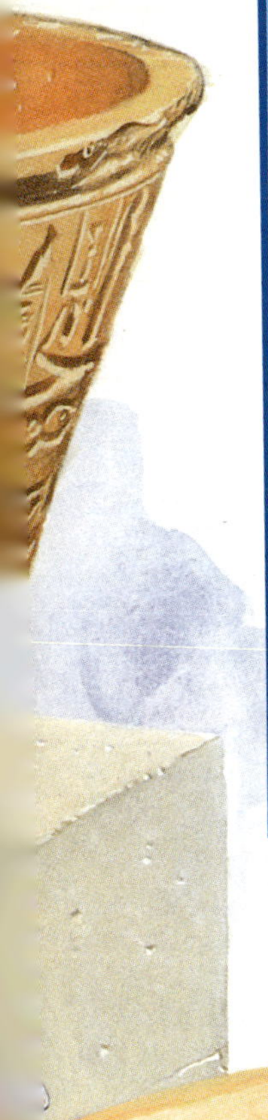

Research Other ancient "clocks" included sand timers and round sundials. Find out how and where these were used.

A fragment of a water clock from ancient Egypt.

NUMBER PICTURES

The ancient Egyptians wrote in pictures, called *hieroglyphs*. Some of the hieroglyphs represented numbers.

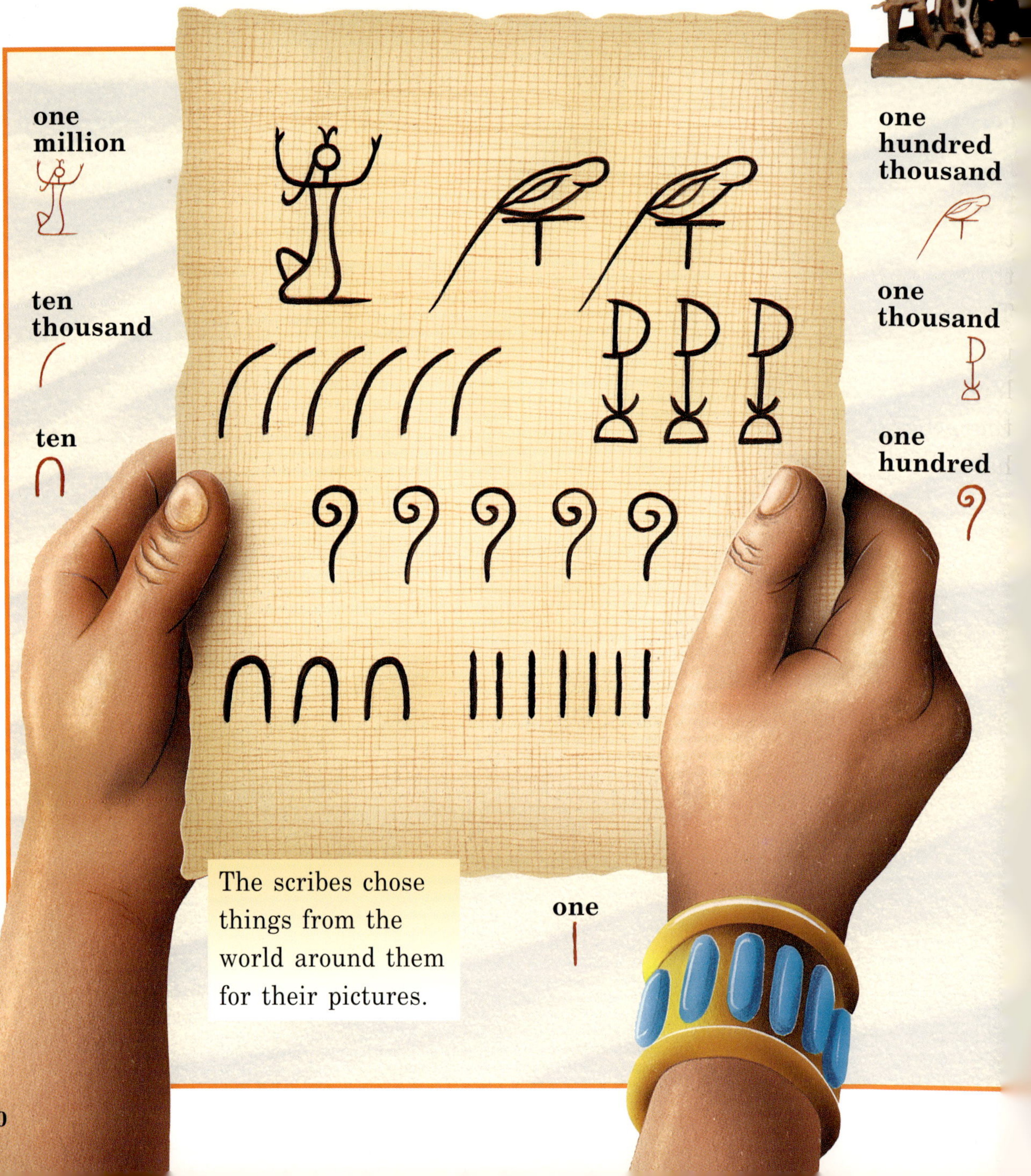

The scribes chose things from the world around them for their pictures.

Counting the Animals

This model, found in a tomb, shows a person's cattle being counted. The painting below shows geese being counted.

One of the important jobs in ancient Egypt was keeping track of the number of animals that people owned. This information was used in calculating how much tax the owner had to pay.

The Egyptians invented papyrus, a kind of paper made from the papyrus plant. Many records were kept on papyrus scrolls.

A Papyrus Problem

One famous Egyptian scroll, known as the Rhind Papyrus, is 18 yards long – and it is full of mathematical problems and calculations. One calculation is still a mystery. It may be a kind of number puzzle.

Look at the illustration of the papyrus problem.

1. a. There are five numbers written above the line. What are they?

b. How does each number compare with the one before it? What pattern do you notice?

c. Add the five numbers. What is the total?

2. What number is written below the line?

3. Use Egyptian number symbols to write an addition problem for a friend to solve.

Problem Number 79 from the Rhind Papyrus

Did you know?
The numbers in the papyrus problem are similar to those in a famous word puzzle:

As I was going to Saint Ives,
I met a man with seven wives.
Every wife had seven sacks.
Every sack had seven cats.
Every cat had seven kits.
Kits, cats, sacks, and wives –
How many were going to
Saint Ives?

What do you think the answer is?

Research

- Find out how the Romans wrote numbers. Make a list of the symbols they used and their values.

Ancient Art

The Egyptians were lovers of beautiful things. They made a wide range of carefully crafted ornaments, jewelry, and pottery, and often decorated objects with bright colors and detailed designs.

Tutankhamun's tomb was filled with decorative objects, such as this gold and jeweled chest piece.

Many Egyptian designs show balance or *symmetry.* This was often a feature of designs based on flower and leaf shapes.

carved amulet, vorn as a kind f necklace.

A painted vooden case r eye "make-up."

Three designs used on head-band pieces.

The Egyptians sometimes used "over and under" designs.

Repeating Patterns

Many Egyptian temples and tombs were decorated with border designs. These were created by repeating an image at equal distances along a line. One element of the design could be used as a stencil.

Many Egyptian designs used repeating patterns of flowers.

1. Look at the repeating pattern above. Suppose you wanted to continue the pattern using a stencil. How far would you slide the stencil each time?
2. In what ways does the gold hawk on page 24 show symmetry? What parts are *not* symmetrical?
3. Select an example of a symmetrical design from page 25. Describe how it shows symmetry.
4. In what ways are the headband pieces on page 25 all similar?

Create Your Own Design

You will need:

a piece of heavy paper, at least 4 x 4 inches; a long piece of construction paper, at least 6 x 18 inches; scissors; ruler

Step 1: Fold the heavy paper in half. Then draw a shape along the folded edge.

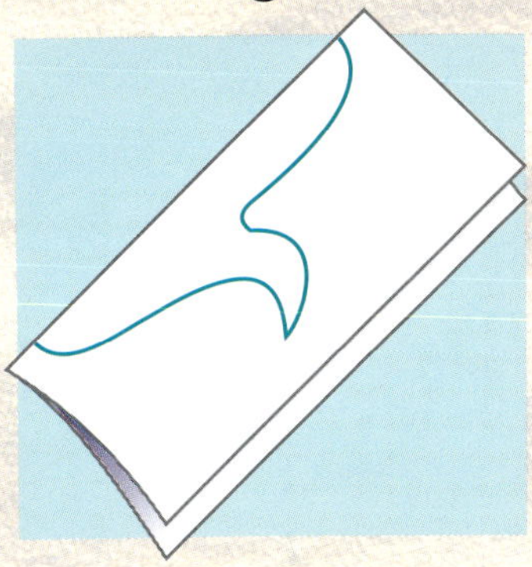

Step 2: Cut out your design and unfold the paper.

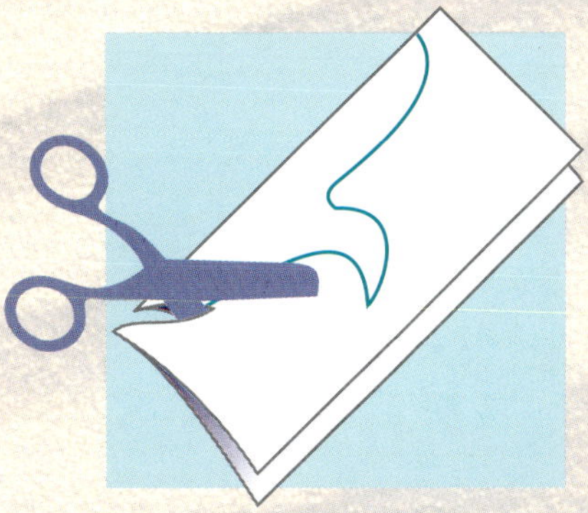

Step 3: Trace the cutout onto the construction paper. Then draw parallel lines to mark the borders of your design.

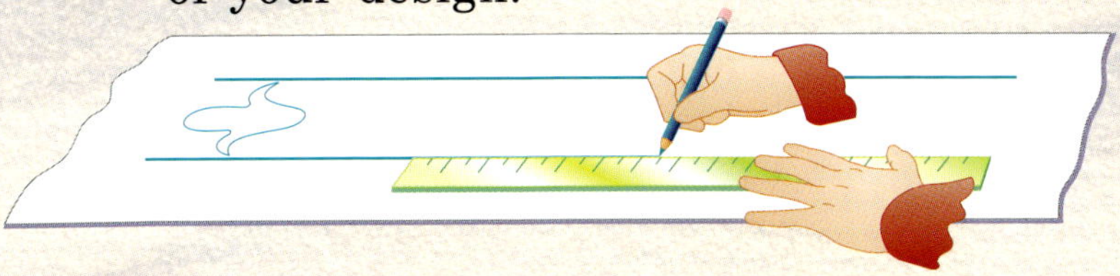

Step 4: Slide the cutout along and trace it again. Continue this repeating pattern, making sure the images are the same distance apart. Color your completed design.

Measuring Up

The Egyptians were the first people to develop precise methods of measurement. When they needed to measure length, they could do so accurately whether the distance was the length of a finger, or the length of a field.

Digits, Palms, and Cubits

Like people from many cultures, the ancient Egyptians used parts of the body as the basis for units of length. There were three main units: the "digit," the "palm," and the "cubit."

DIGIT Width of a finger.

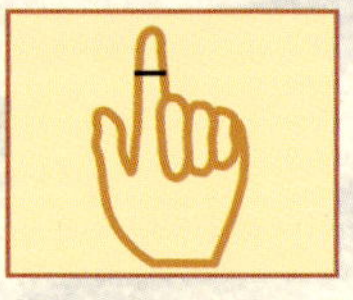

PALM Width of the palm of the hand.

1 palm = 4 digits

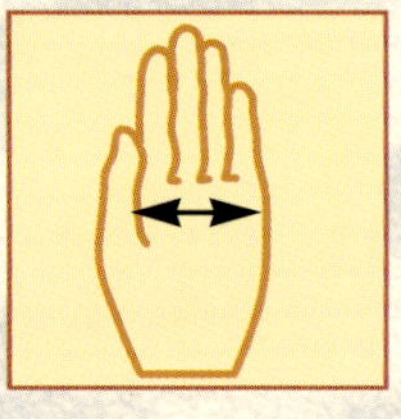

CUBIT This was based on the distance from elbow to fingertip. The Egyptians set a standard length for the cubit; it was equal to about 21 of today's inches.

1 cubit = 7 palms

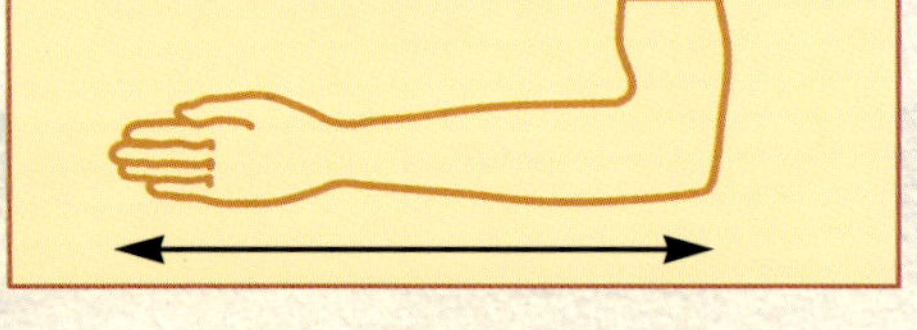

The first unit

The cubit is the first unit of measurement tha people are known to have used. It was described in an Egyptian papyrus more tha 3,500 years ago.

Cubit Ruler
A cubit measuring rod, with markings showing a cubit divided into 28 “digits.”

For some measuring tasks, the Egyptians used lengths of knotted cord. The knots were evenly spaced (for example, one cubit apart) so that distance could be measured by counting the sections of the cord.

The man in this painting is using a knotted measuring cord. He could be measuring the boundaries of a farm for taxation records.

Tools of the Trade

The ancient Egyptians did not only measure length. They also had methods for measuring angles, and they developed ways of measuring weight accurately.

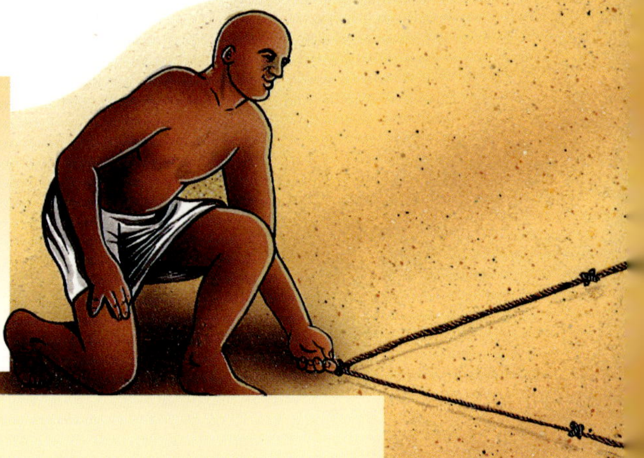

The Egyptians found a way of using a knotted cord to make a square corner. When they formed a triangle with sides 3, 4, and 5 sections long, the corner between the two shorter sides made a right angle.

1. Make a chain of 24 links.

a. Draw all the triangles you can make. Record the length of the sides.

b. Which triangle has a right angle? Mark the square corner.

Look at the units of length shown on page 28.

2. Measure the length of your cubit:

a. in palms and digits **b.** in inches.

How do these measurements compare with the Egyptian cubit?

3. What fraction of a cubit is:

a. one digit? **b.** one palm?

4. What might the ancient Egyptians have measured using:

a. palms? **b.** cubits?

Research

Another early device made from knotted cord was the "quipu," developed by the Incan people of Peru. Find out what a quipu looked like and how it was used.

Weight

A 5-deben weight

Weight was measured in units called "debens" and "kites." One deben was equal to 10 kites. Sometimes weights were made in the shape of animals. This tomb painting shows a bull-shaped weight on a large pan balance.

FUN AND GAMES

The ancient Egyptians developed a wide variety of games. Many "boards" used for games have been found in Egyptian tombs, together with ivory or wooden counters.

Hounds and Jackals

Nobody knows what this race game was called originally. Ever since it was found in a tomb, it has been called "Hounds and Jackals" after the five hound pegs and five jackal pegs that were used as counters.

Senet

"Senet" was a game played by adults. The rectangular board was divided into squares.

The Egyptians played a kind of checkers as long ago as 1300 B.C., using a "board" and counters made of stone.

A wooden and ivory Senet board from the tomb of pharaoh Tutankhamun.

MAKE A MOVE

Nine Men's Morris is the modern name for one of the games invented in ancient Egypt. You can play it using the board shown on page 35; you just need a partner and some counters. For some other games, the Egyptians also needed "throw sticks"; these were used to determine how many moves to make.

Throw Sticks

One side of each throw stick was white, and the other side was black. A player threw four sticks.

Throws and scores

One white side up	1
Two white/two black	2
Three white/one black	3
Four white	4
Four black	6

1. Color four craft sticks to make your own set of throw sticks. Throw the four sticks 20 times. Tally the results. Describe what you notice about the scores.

2. The Egyptian throw sticks do not let you score 5. Make a set of throw sticks that will let you score 1, 2, 3, 4, 5, or 6. List the possible throws and the score for each one.

A Morris Board

How to Play

This is a game for two players. Each player needs nine counters – a different color for each player.

1. The players take turns placing a counter on one of the 24 dots, until all counters are on the board.

2. When all the counters are in place, the players take turns moving a counter to a neighboring vacant dot. ("Jumping" a dot is not allowed.) The aim is to make a row of three counters. This is called a "mill."

3. If a player makes a mill, he or she removes one of the opponent's counters – as long as that counter is not part of a mill.

4. A player wins the game when the opponent is unable to move or has only two counters remaining on the board.

Creative Calculations

The ancient Egyptians had methods of multiplying and dividing that were very different from those we use today. When they needed to multiply or divide, they used a system of doubling.

Did you know?
There are many ways of multiplying numbers. The Egyptian method is the oldest method ever recorded and is still used in some cultures.

A Multiplication Example

Here are the steps for multiplying 13 by 6 using the Egyptian method.

First factor		Second factor
13	x	6

1. Write the number 1 and the second factor.

1	6

2. Start doubling the numbers in each column. Stop before the *first* column reaches a number greater than the first factor.

1	6
2	12
4	24
8	48

3. Choose numbers from the *first* column that you can add together to get the first factor.

1
4
+8
13

1	6
2	12
4	24
8	48

6
24
+48
78

4. From the second column, choose the "partners" to these numbers. Adding these numbers gives you the answer.

SO

13 x 6 = 78

A Division Example

Similar steps were followed for division. Here are the steps for dividing 144 by 6.

Dividend		Divisor
144	÷	6

1. Write the number 1 and the divisor.

1	6

2. Start doubling the numbers. Stop before the *second* column reaches a number greater than the dividend.

1	6
2	12
4	24
8	48
16	96

3. Choose numbers from the *second* column that you can add to get the dividend.

4. From the first column, choose the "partners" to these numbers. Adding these numbers gives the answer.

1	6
2	12
4	24
8	48
16	96

48
+96
144

8
+16
24

SO

144 ÷ 6 = 24

LESS THAN ONE

Measurements often involve numbers that are less than one. The ancient Egyptians developed two creative ways of recording fractions.

The Egyptians used *hekats* and *khars* to measure grain. One khar "sack" was equal to 16 hekats.

The Eye of Horus

For measurements of grain, the Egyptians used the series of fractions created by starting at 1, and then halving. (This gave $\frac{1}{2}$, $\frac{1}{4}$, $\frac{1}{8}$, $\frac{1}{16}$, $\frac{1}{32}$, $\frac{1}{64}$, and so on.)

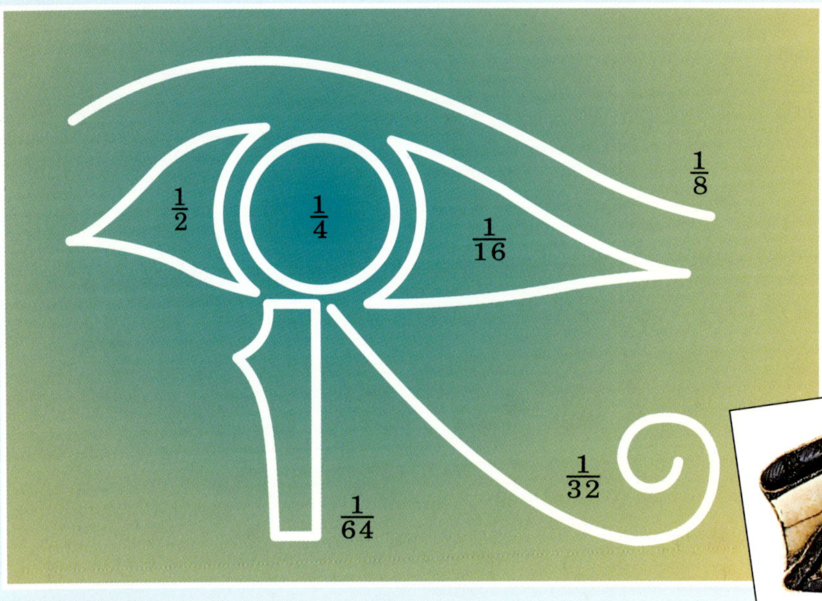

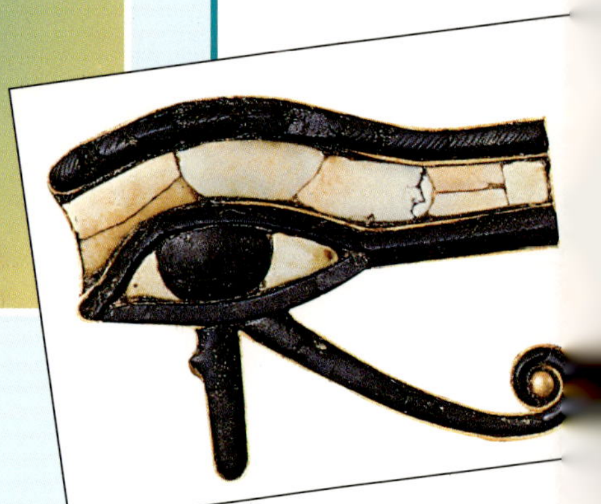

A famous image, *The Eye of Horus,* gave the scribes a "short cut" for writing these fractions. For example, drawing  was a way of showing $\frac{1}{32}$.

"One Part Of"

Another way of showing fractions was to use a special symbol that meant "one part of."

To show "one part of 32" or $\frac{1}{32}$, the Egyptians could write:

1. These jars have their capacity marked on the side. What fractions are shown?

2. Write these fractions using the "one part of" symbol.

a. $\frac{1}{10}$ **b.** $\frac{1}{5}$ **c.** $\frac{1}{20}$

3. If the first basket below holds 1 khar of grain, about how much would each of the other baskets hold? Write your answers using symbols from the eye of Horus.

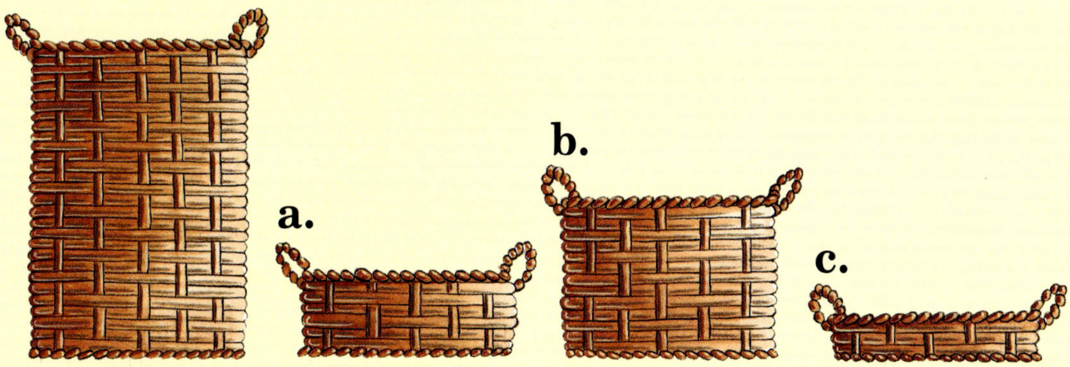

4. One hekat is 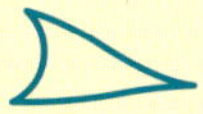 of a khar. Use symbols from the eye of Horus to write the following as fractions of a khar:

a. 8 hekats **b.** 4 hekats **c.** 2 hekats.

GLOSSARY

Angle
Two straight lines form an *angle* when they join at a corner. A *right angle* is an angle that makes a square corner.

Base
The bottom of a shape. The base of a three-dimensional shape is the face that it sits on.

Dividend
In a division problem, the *dividend* is the number that is divided into equal parts.

Divisor
The number in a division problem that a dividend is divided by.

Edge
A line where two faces of a three-dimensional shape meet.

Faces
The flat surfaces of a three-dimensional shape.

Factor
A whole number that can divide a number exactly is a *factor* of that number. For example, 3 is a factor of 12.

Grid
A network of lines forming squares.

Perimeter
The distance around the outside of a two-dimensional shape.

Pyramid
A three-dimensional shape with triangular faces that form a point at the top. A pyramid's base can be any straight-sided, flat shape.

Pyramidal number
The number of objects that can be stacked to make a pyramid shape on a square base.

Slide
To move a shape or figure in a straight line to make a repeating pattern.

Symmetry
When one side of a shape is a mirror image of the other side, the shape is said to be *symmetrical*. Objects can have other kinds of symmetry, too For example, if a shap can be turned through a fraction of a full rotation and still look the same, it is said to have *turning symmetry*

Vertex
A point, or corner, where the edges of a shape meet. The word for more than one vertex is *vertices*. For example, a triangle ha three vertices.

INDEX